HYDROPONICS

Explore the basics of hydroponics, a method of growing plants without soil.

Rita van Klaveren

Table of Contents

HYDROPONICS 1

INTRODUCTION...4

CHAPTER 1 8

Introduction to Hydroponic system.............................8

How it works 13

Components of hydroponic system 15

CHAPTER 2...20

Types Of Hydroponic Systems: Advantages and

Disadvantages..20

CHAPTER 3..54

Benefits of Hydroponics..54

CHAPTER 4..64

Choosing Hydroponic System For Your Needs.........64

CHAPTER 5..71

Plant Selection in Hydroponics 71

Factors to consider 71

CHAPTER 6..81

Types of Nutrients and Nutrient Solutions Used in Hydroponics 81

Essential Nutrients for Hydroponics 82

Common hydroponics nutrient solutions 84

CHAPTER 7..90

Troubleshooting Common Hydroponic Issues:

Solutions for Optimal Growth.....................................90

CHAPTER 8...102

Future of Hydroponics 102

Acknowledgement..108

ABOUT THE AUTHOR...109

INTRODUCTION

Hydroponics is an intriguing way of growing plants without soil. Instead, plants are grown in a nutrient-rich water solution, resulting in faster development and larger harvests. It's an excellent choice for beginners who wish to grow their own plants inside or in limited places, as it takes up less space and can be easier to handle than traditional soil-based gardening. This book will provide you with all the knowledge necessary to start hydroponic gardening, including information on various types of hydroponic systems, the best plants for hydroponic growth, and how to maintain balanced nutrient levels for your plants. With the proper expertise and equipment, you can quickly reap the benefits of hydroponic farming. Hydroponics is a way of growing plants that does not require soil. It has gained popularity because it can help conserve water and provide higher yields in a controlled setting. This method uses a nutrient-rich solution to supply important minerals and nutrients straight to the plant roots, resulting in faster development and healthier plants. Both commercial agriculture and home gardening have utilized this method, providing a more efficient and sustainable way to grow a wide range of crops.

Yes, hydroponics may be the way to go if you're interested in producing your own plants without using soil.

Hydroponics is an excellent method for producing healthy and sustainable crops since it allows for control over nutrient levels, water usage, and plant growth.

Are you prepared to transform your growth process? Say farewell to conventional soil-based farming and welcome the amazing realm of hydroponics!

Participate in the hydroponic revolution now and observe the imminent advancement of agriculture unfold right in front of you. Cultivate the principles of sustainability and reap the benefits of a more environmentally friendly future with this guide!

CHAPTER 1

Introduction to Hydroponic system

Hydroponics, an approach to cultivating plants without the use of soil, has garnered significant interest in recent times due to its manifold advantages and capacity to fundamentally transform contemporary agriculture. Unlike traditional agricultural methods, which rely on soil for plant growth, hydroponics uses nutrient-rich water solutions as the principal growing medium. This novel technology feeds plants with precisely regulated nutrients, water, and illumination in a controlled environment, resulting in faster growth, larger yields, and more sustainable agricultural practices.Hydroponics involves growing plants in various ways that allow their roots to come into close contact with the nutrient solution. These systems include simple setups like nutrient film technique (NFT) and deep water culture (DWC), as well as more advanced designs like aeroponics and vertical farming. Hydroponic gardening has various benefits over conventional soil-based farming, regardless of the type of system employed. To begin with, hydroponics facilitates the most efficient use of resources, particularly water. In contrast to conventional agricultural practices that frequently experience water wastage as a result of evaporation and inadequate soil absorption, hydroponic systems facilitate water recycling and conservation.

Hydroponics can reduce water usage by up to 90% through constant recirculation of the nutrient solution, making it an environmentally sustainable option in areas with low water supplies or drought situations. Furthermore, hydroponic systems provide exact control over nutrient content, ensuring that plants receive the optimal blend of minerals and vitamins for growth. Compared to crops cultivated normally, these customized plants are often more vibrant, healthier, and have a higher nutritional content. Hydroponics, with its capacity to adjust nutrient solutions to individual plant demands, allows producers to produce high-quality, nutritious foods.Versatility in terms of location and space utilization is another important benefit of hydroponics.

Hydroponic gardening is beneficial in urban areas with limited land availability since it allows for vertical farming and maximizes space efficiency. Hydroponics allows for urban farming by exploiting abandoned buildings, roofs, and underutilized urban areas, increasing local food production while minimizing the carbon footprint associated with long-distance transportation from rural farms.

Hydroponic farming not only offers these advantages but also lowers the risk of soil-borne diseases and pests that are often associated with traditional farming. Hydroponics reduces the risk of infection and does away with the need for chemical pesticides and fertilizers, resulting in cleaner and safer produce by cultivating plants in a sterile and regulated environment. This is why consumers who are concerned about their health and want to reduce their exposure to dangerous chemicals may find hydroponically grown fruits and vegetables to be an appealing option. All things considered, hydroponics provides a productive and sustainable substitute for conventional soil-based agriculture. Hydroponics has the potential to revolutionize the way we grow food because of its reduced environmental impact, enhanced crop yields, customized fertilizer control, and water efficiency. In light of global issues such as population growth, shrinking arable land, and climate change, hydroponics offers a viable way to guarantee food security, support regional farming, and preserve finite resources.

How it works

Hydroponic systems function by providing precise control over environmental parameters such as temperature and pH balance, as well as increased exposure to nutrients and water. The basic idea behind hydroponics is to provide plants exactly what they require at the right time. Hydroponic systems apply nutrient solutions specifically designed for the needs of the plant being produced. They provide you with complete control over the amount and duration of light that the plants get. It is possible to monitor and modify pH levels. Highly tailored and regulated environments accelerate plant growth.Managing the plant's surroundings decreases numerous risk factors. Plants cultivated in fields and gardens are exposed to numerous factors that detrimentally affect their health and growth. Plant diseases can be spread by fungi found in the soil. Raccoons and other wildlife have the ability to steal ripe veggies from your garden. Locusts and other pests can destroy crops in the course of an afternoon. Hydroponic systems eliminate the unpredictable nature of growing plants in the ground and outdoors. Removing the soil's mechanical barrier allows seedlings to mature significantly faster. Hydroponics produces significantly healthier, higher-quality fruits and vegetables by doing away with pesticides. Plants are free to develop quickly and strongly in the absence of impediments.

Components of hydroponic system

A hydroponic system's key components include a nutrient solution that provides essential nutrients for plant growth, a growing medium that supports the plant roots and helps with nutrient absorption, a water supply that provides a constant flow of water to the plants, an aeration system that ensures proper oxygen levels in the nutrient solution, lighting that provides the necessary light for photosynthesis, and pH control to maintain the proper levels for optimal plant growth. In order to keep your hydroponic system flourishing, you'll need to familiarize yourself with a few parts that keep hydroponics functioning smoothly.

- **Growing media**

Hydroponic plants are cultivated in inert media that serve to stabilize the plant's root system and support its weight. Although growing media serves as an alternative to soil, it does not supply the plant with any supplementary nutrients. Instead, this porous medium stores moisture and nutrients from the nutrient solution, which it subsequently transfers to the plant. Furthermore, a lot of growing media are pH-neutral, so they won't throw off the equilibrium of your nutrient solution. A variety of media are available; the one that works best for your project will depend on the particular plant and hydroponic system. There are several places to get hydroponic growth medium, including local nurseries and gardening stores as well as online.

- **Air stones and air pumps**

If the water is not adequately aerated, submerged plants have a high risk of rapidly drowning. The presence of air stones distributes small bubbles of dissolved oxygen throughout the reservoir of your nutritional solution. These bubbles also contribute to the solution's ability to distribute the dissolved nutrients in an even manner. It is not possible for air stones to produce oxygen on their own. Because they are opaque, air stones require the use of food-grade plastic tubing to connect them to an external air pump, preventing the growth of algae. This will prevent the growth of algae from taking place. Air pumps and air stones are two of the most common components found in aquariums, and they are readily available for purchase at pet stores.

- **Net pots**

Hydroponic plants are grown using net pots, which are mesh planters. The latticed structure of net pots exposes roots to more oxygen and nutrients as they grow out of the pot's bottom and sides. When it comes to drainage, net pots outperform conventional clay or plastic pots.Essential parts of a hydroponic system include an aeration system, water and nutrient solution, growing media, pump, and reservoir. In order to create the perfect atmosphere for plants to flourish without soil, each of these elements is essential. It is possible to set up and maintain an effective hydroponic system for your plants if you comprehend how these parts interact. You may take advantage of the advantages of hydroponic plant growth, including robust, vibrant plants, with the right maintenance and attention to these elements.

CHAPTER 2

Types Of Hydroponic Systems: Advantages and Disadvantages

It's critical to take into account the various hydroponic system types when designing an indoor or outdoor garden. There are several systems to learn about and comprehend how they operate, ranging from deep water culture to nutrient film technique, to assist you in selecting the most suitable one for your requirements.

- **Wick systems**

Growers raise plants in wick systems by placing them on a tray above a reservoir. This reservoir contains a nutrient-dissolved water solution. Wicks move from the reservoir to the growing tray. The growth medium surrounding the plant root systems becomes saturated with water and nutrients as they travel up the wick. You can make these wicks using simple materials like felt, rope, or string. Wick systems are the most basic type of hydroponics available. Wick systems are hydroponics that don't require mechanical components, like pumps, to work. It is therefore perfect in scenarios where electricity is either unavailable or inconsistent.Wicks systems function through a mechanism known as capillary action. The wick absorbs the water it is submerged in, transferring the nutritional solution to the porous growing medium, similar to how a sponge works. Wick system hydroponics can only be successful in the presence of growing media that can promote the transfer of water and nutrients. Fibers from the outer husks of coconuts, known as coco coir, have the extra advantage of being pH neutral and having great moisture retention. In addition to being highly porous and pH neutral, perlite is perfect for wicking systems. In addition to being highly porous, vermiculite has a high capacity for cation exchange. It can thus store nutrients for use at a later time. The best growing media for hydroponic wick systems are these three.Wick systems are less useful for growing larger plants because they operate more slowly than other hydroponic systems. Make sure that each plant in the growth tray has a minimum of one wick extending from the reservoir. Position these wicks in close proximity to the plant's root system. Even though the wick system can operate with aeration, many users still decide to add an air stone and air pump to the reservoir. This increases the hydroponic system's oxygenation level.

Advantages of a wick system

- *Simplicity:* A wick system is simple to set up and requires little maintenance once operational. The wicks will constantly feed water to your plants, eliminating the possibility of them drying out. Also, plants like lettuce will thrive in a wick system, delivering a high return on your hands-free investment.

- *Space-efficient:* Since wick systems don't require

power to operate, they are discreet and may be placed anywhere. For educators, novices, or anybody curious about hydroponics, this setup is ideal.

Disadvantages of a wick system

- *Limitations:* Herbs like basil, mint, and rosemary, as well as lettuce, grow quickly and don't require a lot of water. However, due to their high nutrient and water requirements, tomatoes will not do well in a wick system. Constant moisture is not conducive to the growth of other plant species. Carrots and turnips are examples of root vegetables that will not thrive in a wick system.

- *Susceptible to rot:* A hydroponic wick system is consistently humid and moist. As a result, your plants' roots and the organic growing medium are vulnerable to fungal outbreaks and rot.

- **Deep water culture systems**

Deep water culture hydroponics are basically plants hung in oxygenated water. Deep water culture systems, or DWC systems for short, are among the simplest and most widely used hydroponics techniques available. In a DWC system, net pots containing plants are positioned below a suspended deep reservoir of an oxygen-rich nutrient solution. The solution provides constant access to nutrients, water, and oxygen for the plant's roots. Some people believe that deep water culture is the purest type of hydroponics. Water oxygenation is essential to the life of plants since their root systems are constantly submerged in water. If the plant's roots do not receive sufficient oxygen, it will drown in the solution. Add an air stone connected to an air pump at the reservoir's bottom to oxygenate the entire system. Additionally, the air stone's bubbles will aid in the nutrient solution's circulation.A deep water culture system can be easily assembled in a classroom or at home without the need for expensive hydroponics equipment. To hold the solution, use a clean bucket or old aquarium and cover it with a floating surface such as styrofoam to house the net pots. Immerse only the roots of plants in DWC systems in the solution. There should be no submerged portions of the stem or vegetation. Even approximately one and a half inches of roots should remain above the waterline. The roots won't be in danger of drying out since the air stone bubbles will pop through the surface and splash onto the exposed roots.

Advantages of Deep Water Culture Systems

- *Low maintenance:* Once configured, a DWC system requires very minimal maintenance. Just make sure your pump is running oxygen to the air stone and refresh the nutrient solution as needed. The nutrition solution is normally replenished every 2–3 weeks, but this varies depending on the size of the plants.

- *DIY appeal:* Unlike many hydroponic systems, deep water culture systems can be easily and economically assembled at home by purchasing the necessary nutrients and air pump from a nearby nursery or pet store.

Disadvantages of Deep Water Culture Systems

- *Limitations:* Larger and slower-growing plants are difficult for deep water culture systems to grow, although they do well with lettuce and herbs. DWC systems are suboptimal for flowering plants. Nevertheless, with additional effort, it is possible to cultivate plants such as tomatoes, bell peppers, and squash in a DWC system.

- *Temperature control:* It's crucial that the temperature of your water solution stay between 60°F and 68°F. Because the water in a DWC system is static and not recirculating, temperature control may be more challenging.

- **Ebb and flow systems**

Ebb and flow hydroponics operates by flooding a grow bed with a nutrient solution from a reservoir beneath. There is a timer on the submersible pump in the reservoir. Upon activation of the timer, the pump transfers the water and nutrients to the grow bed. Gravity gradually removes the water from the grow bed and returns it to the reservoir when the timer stops. The system prevents floods from reaching a specific level and harming the plant stalks and fruits by using an overflow tube. Ebb and flow systems differ from other methods because the plants do not remain continuously submerged in water. The plants absorb the nutrient solution through their roots while the grow bed is flooded. The roots dry out when the grow bed empties and the water level drops. Then, in the void before the subsequent flood, the dry roots absorb oxygen. The size of your grow bed and the size of your plants will determine how long it takes between floods.One of the most often used hydroponic gardening techniques is the use of ebb and flow systems, commonly known as flood and drain systems. The plants develop quickly and vigorously because they receive a lot of oxygen and nutrients. The ebb and flow mechanism is highly adaptable and easily adjustable. You can place a variety of fruits and vegetables, along with net pots, inside the grow bed. The ebb and flow system gives you the most flexibility to experiment with your plants and mediums in any hydroponic system.Ebb and flow systems can accommodate almost any kind of vegetation. The grow tray's dimensions and depth are your main constraints. Compared to lettuce or strawberries, root veggies will need a considerably deeper bed. Among the most common ebb and flow crops are tomatoes, peas, beans, cucumbers, carrots, and peppers. You can fasten trellises directly to the grow bed. In ebb and flow hydroponics, "grow rocks" and enlarged clay pebbles (hydroton) are among the most widely used growing media. These are lightweight, reusable, and cleanable. Although they do hold moisture, they also drain. This is a crucial feature of flow and ebb systems.

Advantages of an ebb and flow system

- *Versatility:* An ebb and flow system allows for larger plant growth compared to most conventional hydroponic setups. Ebb and flow hydroponics works incredibly well for producing fruits, flowers, and veggies equally. You'll get an abundant output if you've taken care to give your plants the right size grow bed and nutrition.

- *DIY appeal:* You may build an ebb and flow hydroponic system at home in countless ways. You can get all the materials you need to build an ebb and flow system by going to hardware and pet stores. Ebb and flow systems are more costly to set up than other do-it-yourself systems like wick and deep water culture, but they support a far wider variety of plant life than they can.

Disadvantages of an ebb and flow system

- *Pump failure:* Like any pump-dependent hydroponic system, your plants will perish if the pump fails to function. It is your responsibility to monitor the performance of your ebb and flow system to ensure that it does not negatively impact the health of your vegetation. Your plants will not receive sufficient nutrients and water if the water is coming and going at an excessive rate.

- *Rot & disease:* An ebb and flow system requires regular maintenance and sanitation. Rot and root infections may develop in the bed if it is not draining adequately. Mold can grow, and insects can be drawn to an unclean ebb and flow system. Neglecting sanitation leads to crop damage. The extremes in flooding and draining can cause an abrupt pH shift that is unsuitable for some plants.

- **Nutrient film technique systems**

Nutrient film technique (NFT) systems suspend plants above a stream of continuously flowing nutrient solution th Nutrient film technique (NFT) systems hold plants above a constant stream of nutrient solution that washes over the ends of the root systems. The tubes tilt to allow water to flow down the grow tray and into the reservoir below. Then, an air stone aerates the water in the reservoir. A submersible pump then removes the nutrient-rich water from the reservoir and returns it to the channel's surface. A recirculating hydroponic system is what the nutrient film technique is.In contrast to deep water culture hydroponics, an NFT system does not submerge the plant roots in water. Rather, the "film" or stream merely extends beyond the tips of their roots. The exposed root system receives plenty of oxygen, and the tips of the roots will draw moisture up into the plant. The grooved bottoms of the channels allow the shallow film to easily slide over the root tips. Additionally, by doing this, water cannot accumulate or block up against the root systems.Although nutrient film technology systems recycle water continuously, it is a good idea to empty the reservoir and add fresh nutrient solution approximately once a week. This guarantees that your plants are receiving enough nutrients. You must slope NFT channels gradually to ensure adequate nourishment for the plants. Water will rush down the channel if it is excessively steep, failing to adequately nourish the plants. If an excessive amount of water is forced through the channel, the system may overflow, potentially drowning the plants. Commercial settings widely use nutrient film technique (NFT) hydroponics systems because they are easy to mass-produce and can sustain multiple plants per channel, making them ideal for lightweight plants like strawberries and lettuce, spinach, mustard greens, and kale. Fruiting plants with higher yields, such as tomatoes and cucumbers, will need trellises to support their weight.

 at washes over the ends of the plant's root systems. The channels holding the plants are tilted, allowing water to run down the length of the grow tray before draining into the reservoir below. The water in the reservoir is then aerated via air stone. A submersible pump then pumps the nutrient-rich water out of the reservoir and back to the top of the channel. The nutrient film technique is a recirculating hydroponic system.

Unlike with deep water culture hydroponics, the roots of the plants in an NFT system are not immersed in water. Instead, the stream (or "film") only flows over the ends of their roots. The roots' tips will wick the moisture up into the plant, while the exposed root system is given plenty of access to oxygen. The bottoms of the channels are grooved, so the shallow film can pass over the root tips with ease. This also prevents water from pooling or damming up against the root systems.

Even though nutrient film technique systems are constantly recycling water, it is wise to drain the reservoir and replenish the nutrient solution every week or so. This ensures your plants are being delivered ample nutrition. NFT channels must be angled at a gradual slope. If it's too steep, the water will rush down the channel without properly nourishing the plants. If too much water is being pumped through the channel, the system will overflow and the plants can drown. NFT hydroponics are popular commercial systems, as they can support several plants per channel and can easily be mass-produced. Nutrient film technique systems are best suited for lightweight plants, like mustard greens, kale, lettuce, spinach as well as fruits like strawberries. Heavier fruiting plants like tomatoes and cucumbers will require trellises to support the excess weight.

Advantages of a nutrient film technique system

- Low consumption: NFT hydroponics don't require a lot of water or nutrients to work because they recirculate the water. Salts find it more difficult to build up on the roots

of the plant due to the continuous flow. Additionally, because nutrient film technique systems don't require growing media, you can avoid the cost and inconvenience of buying and replacing them.

- Modular design: Systems using the nutrient film technique are ideal for commercial and large-scale projects. Expanding a channel is relatively simple once it is up and running. You can install several channels in your greenhouse to support various crops. Using a different reservoir for each channel is a smart idea. In this manner, you won't lose your entire enterprise in the event of a pump failure or the spread of disease in the water.

Disadvantages of a nutrient film technique system

- *Pump failure:* If the pump breaks down and the nutrient film is not circulated through the channel, your plants will dry out. If your crop isn't getting enough water, it could all die within a few hours. An NFT hydroponic system does need careful maintenance. You should keep a close eye on how well your pump is operating.

- *Overcrowding:* If the plants are placed too closely together or if the root development is excessive, they may clog the channel. Roots blocking a channel will prevent water from getting to the plants, which will lead to their starvation. This particularly applies to the plants at the channel's bottom. Take into consideration eliminating some plants or converting to a smaller unit if the plants at the end ever seem to be underperforming in comparison to the remainder of the channel.

- **Drip systems**

An aerated and nutrient-rich reservoir distributes solution to individual plants via a network of tubes in a hydroponic drip system. The drip system slowly drips this solution into the growing substrate surrounding the root system, keeping the plants hydrated and nourished. Drip systems are the most common type of hydroponics, particularly among commercial producers. Individual plants or large-scale irrigation operations can use drip systems.Drip system hydroponics comes in two configurations: recovery and non-recovery. Recovery systems, which are more common among smaller home growers, recirculate extra water by draining it from the grow bed back into the reservoir during the subsequent drip cycle. The extra water in non-recovery systems flows off and wastes away from the growing media. It is more common for commercial producers to use this strategy. Despite the seeming wastefulness of non-recovery drip systems, large-scale plants use very little water. These drip systems accurately supply the amount of solution needed to maintain the moisture content of the growing substrate surrounding the plant. Non-recovery drip systems minimize waste by using complex timers and feeding schedules.If you are cultivating plants using a recovery drip system, you must be aware of the variations in the nutrient solution's pH. Any system that allows wastewater to cycle back into the reservoir will experience this. The grower will need to check and adjust the solution reservoir more frequently than they would in a non-recovery system since plants will deplete the solution's nutrient content and change its pH balance. Growing mediums will need to be cleaned and replaced on a regular basis because they can also become oversaturated with nutrients.

Advantages of a drip system

- Variety of plant options: A drip system can support larger plants better than the majority of other hydroponic

systems. Commercial growers find it very enticing for this reason. An appropriately sized drip system can provide ample support for melons, pumpkins, onions, and zucchini. Because drip systems may accommodate larger root systems, they can carry more growth medium than other types of systems. Peat moss, rockwool, and coco coir are examples of slow-draining media that are ideal for drip systems.

- *Scale:* Drip systems are highly capable of facilitating extensive hydroponics operations. To increase the number of plants, a grower can connect additional tubing to a reservoir and redirect the solution to the extra vegetation. Growers can incorporate additional reservoirs with customized timer schedules into an existing drip system to accommodate the requirements of newly introduced crops. Another contributing element to the popularity of drip systems in commercial hydroponics is their efficiency.

Disadvantages of a drip system

- Maintenance: If you are cultivating plants using a non-recovery drip system at home, there is a substantial level of upkeep required. To ensure optimal conditions, it is imperative to regularly assess the pH and nutrient levels of your solution, and, if needed, perform drainage and replacement. Debris and plant matter might obstruct the recovery system lines, necessitating regular cleaning and

flushing of the supply lines.

- *Complexity:* Drip systems can readily become complicated and complex projects. However, professional hydroponics may be less affected by this, but it is not the optimal solution for home growers. There are some less complex methods, such as ebb and flow, that are more suitable for hydroponics practiced at home.

- **Aeroponics**

Plants in aeroponics systems are suspended in the air, exposing their bare roots to a nutrient-rich mist. Aeroponics systems are enclosed structures, such as cubes or towers, designed to accommodate several plants simultaneously. The system uses a reservoir to store water and nutrients, which it then pumps to a nozzle. The nozzle transforms the solution into a fine mist and distributes it. The uppermost part of the tower often emits the mist, allowing it to flow down the chamber in a cascading manner. Aeroponics involves the continuous misting of the plant's roots, similar to how NFT systems constantly expose the roots to a nutrient film. Others operate in a manner similar to an ebb and flow system, intermittently showering the roots with a fine mist. Aeroponics can thrive without the need for any substrate media. The root's continuous exposure to air enables it to absorb oxygen and grow rapidly.Aeroponics systems consume a lower amount of water compared to other hydroponic methods. Actually, compared to an irrigated

49

field, aeroponically grown crops require 95% less water. Their vertical configuration is specifically engineered to occupy a limited amount of space and enables the accommodation of multiple towers on a single site. Aeroponics enables the cultivation of high yields in limited places. In addition, due to their enhanced oxygen exposure, aeroponic plants exhibit accelerated growth compared to conventional hydroponically cultivated plants.Aeroponics enables convenient year-round harvesting. Vining plants and members of the nightshade family, such as tomatoes, bell peppers, and eggplants, thrive in an aeroponic setting. Lettuce, baby greens, herbs, watermelons, strawberries, and ginger all thrive as well. Nevertheless, fruiting trees are unsuitable for aeroponic cultivation due to their substantial size and weight, while underground plants such as carrots and potatoes, which possess vast root systems, are not amenable to growth in this method.

Advantages of an aeroponics system

- *Oxygen surplus:* The excess oxygen absorbed by the exposed roots greatly enhances the plant's growth. Aeroponics is not only the most environmentally friendly hydroponic system, but it is also one of the most efficient in terms of performance. These systems are adaptable and can be customized to consistently generate outcomes of superior quality.

- *Mobility:* Aeroponic towers and trays are highly

portable, allowing for seamless plant relocation without any negative impact on their growth. While transporting, it is advisable to spray the roots with a fine mist to avoid dehydration. In addition, aeroponic devices are specifically engineered to be ergonomic and optimize space utilization. Aeroponics enables the cultivation of plants at higher densities compared to traditional hydroponic systems.

Disadvantages of an aeroponics system

- *Expensive*: Aeroponics entails a greater initial expense compared to alternative hydroponic systems. Establishing a comprehensive system equipped with reservoirs, timers, and pumps can incur expenses amounting to thousands of dollars. Constructing a DIY aeroponics system costs less but poses more challenges compared to building a DIY deep water culture or wick system.

- Maintenance: The equilibrium of aeroponics systems is quite sensitive, and any disturbance can have catastrophic consequences for the well-being of your plants. In the event that your timer fails to activate or a pump malfunctions, there is a significant possibility of losing your entire crop unless you manually spray the roots with mist. To maintain the health of your plants, it is essential to frequently clean the root chamber in order to prevent root disease. In general, aeroponic systems necessitate a higher level of technical expertise for successful

implementation compared to other systems.

CHAPTER 3

Benefits of Hydroponics

In recent years, hydroponics, a method of cultivating plants without soil, has become increasingly popular. This novel method entails the growing of plants in nutrient-dense aqueous solutions instead of conventional soil. Hydroponics offers extensive advantages that have a significant impact on both the agricultural sector and global sustainability initiatives. This article outlines the diverse benefits of hydroponics, emphasizing its capacity to transform contemporary agriculture.

- *Increased Crop Yields:*

An important advantage of hydroponics is its capacity to greatly enhance crop productivity. Hydroponics provides a highly regulated and optimized growing environment, in contrast to conventional soil-based farming. Hydroponic systems promote rapid and healthy plant growth by supplying plants with an optimal combination of nutrients, water, and illumination. The optimal use of resources can lead to a potential increase of up to 30% in crop yields when compared to traditional agricultural practices.

- *Conservation of Water Resources:*

Amidst the current period characterized by increasing

water scarcity, hydroponics emerges as a viable and sustainable alternative. Hydroponics, in contrast to traditional agricultural methods, enables the reuse and preservation of water by eliminating the need for excessive amounts of water to compensate for evaporation and inadequate soil absorption. Hydroponic systems, by constantly reusing the nutrient solution, consume up to 90% less water compared to conventional agriculture, rendering them an eco-friendly alternative.

- *Elimination of Soil-Borne Diseases and Pests:*

Hydroponics mitigates the potential for soil-borne diseases and pests, which are prevalent in conventional agriculture. Soil-borne diseases, such as fungus, bacteria, and nematodes, can inflict substantial harm on crops and result in reduced yields. Hydroponics minimizes the risk of infection by cultivating plants in a sterile and regulated environment, enabling farmers to grow crops that are healthy and free from diseases. Amidst the current period of increasing water scarcity, hydroponics offers a viable and sustainable option. Hydroponics, unlike conventional agriculture, enables the reuse and preservation of water by eliminating the need for excessive amounts of water to compensate for evaporation and inadequate soil absorption. Hydroponic systems, with the continual recirculation of the nutrient solution, consume significantly less water, up to 90% less, compared to conventional agriculture. This makes hydroponics an environmentally conscious option.

- ***Increased Nutritional Value of Produce:***

Studies have demonstrated that hydroponically cultivated produce frequently exhibits superior nutritional content in comparison to conventionally grown vegetables. Hydroponic methods ensure plants receive ideal quantities of vital minerals and vitamins by carefully regulating the nutrient composition. By employing this method, the resulting food exhibits elevated amounts of antioxidants, vitamins, and minerals, rendering hydroponic fruits and vegetables a more healthful option for consumers.

- ***Year-Round Crop Production:***

Hydroponics provides the flexibility to grow crops throughout the year, regardless of seasonal constraints. Hydroponics enables growers to manipulate variables like light, temperature, and humidity by establishing an artificial growing environment. This control facilitates the consistent and continual cultivation of crops, minimizing dependence on erratic weather conditions and prolonging the accessibility of fresh, locally cultivated produce.

- *Land Utilization and Urban Farming:*

Hydroponics offers a cutting-edge method to optimize land usage in densely populated urban areas with limited available land. Vertical farming, a hydroponic technique that utilizes stacked layers for plant growth, reduces the required cultivation area. Hydroponics facilitates urban farming by making use of vacant buildings, roofs, and abandoned urban spaces. This practice promotes local food production and helps minimize the negative environmental effects caused by transporting food over large distances.

- *Reduced Dependency on Chemicals and Pesticides:*

Conventional farming frequently depends on the application of chemicals and pesticides to ward off pests and diseases. Nevertheless, these compounds can exert harmful consequences on human health and the environment. Hydroponics mitigates the necessity for these chemicals by eradicating pests found in soil and creating a regulated environment that decreases the likelihood of disease outbreaks. Hydroponic farming effectively decreases the utilization of chemicals, promotes the development of more robust ecosystems, and offers consumers products that are free from pesticides.

- ***Decreased Carbon Footprint:***

Hydroponic gardening has a reduced carbon footprint compared to conventional farming due to its lower water usage and fewer transit requirements from farms to retailers. It aids in reducing greenhouse gas emissions and contributes to a more sustainable food production system.

- *Customizable Nutrient Control:*

Hydroponics allows for modification of the nutrient solution to cater to the individual requirements of each plant, thereby promoting ideal development and absorption of nutrients. This enables meticulous regulation of the nutrient composition, leading to the cultivation of more robust and healthier plants.

Considering the ongoing increase in global population and urbanization, it is imperative to embrace sustainable agriculture practices. Hydroponics has numerous benefits, including enhanced agricultural productivity, efficient water usage, disease avoidance, elevated nutritional content, reduced reliance on chemicals, less environmental impact, and the ability to customize nutrient levels. Hydroponics has the capacity to transform contemporary agriculture by utilizing these advantages, tackling the issues of food security, water shortages, and environmental sustainability. Adopting hydroponics not only encourages sustainable agriculture but also provides consumers worldwide with fresher and healthier crops.

CHAPTER 4

Choosing Hydroponic System For Your Needs

Selecting the appropriate hydroponic system to suit your requirements can be a challenging endeavor, considering the extensive range of choices accessible in the industry. Every system possesses distinct benefits and drawbacks; hence, it is crucial to thoroughly evaluate your requirements and objectives prior to reaching a conclusion. This book will examine the crucial variables to take into account when choosing a hydroponic system, enabling you to make a well-informed decision.

- ***Plant Selection:***

Prior to selecting a hydroponic system, it is essential to ascertain the specific variety of plants you intend to cultivate. Various plants exhibit distinct nutrition and water demands; thus, unique hydroponic systems are more suitable for particular plant species. If you choose to cultivate little herbs or lettuce, an uncomplicated nutrition film technique (NFT) method would be optimal. Conversely, larger plants, such as tomatoes or cucumbers, may necessitate a more intricate drip irrigation system.

- ***Space Availability:***

Take into account the available space for establishing

your hydroponic system. If you have a restricted amount of space available, it may be necessary for you to choose a small or vertical system such as aeroponics or tower gardens. These systems optimize space efficiency by cultivating plants in a vertical configuration. If you have a larger space available, you may consider using a flood and drain (ebb and flow) system or a deep water culture (DWC) system instead. These systems necessitate a larger physical area, yet they provide enhanced capability and adaptability.

- *Cost:*

The cost of hydroponic systems can vary greatly. Establish your financial plan and evaluate what you are capable of purchasing. It is important to consider that the initial expenses associated with establishing a hydroponic system may encompass several components, including grow lights, nutrient solutions, pumps, timers, and pH meters. Additionally, it is important to take into account long-term expenses related to the maintenance and functioning of the system, including electricity usage and the need to replace consumable items.

- *Experience Level:*

When selecting a hydroponic system, it is crucial to take into account your level of expertise in this field. Systems such as drip irrigation or NFT (nutrient film technique) may be more suited for novices because of their simplicity and effortless operation. In contrast, more sophisticated methods such as aeroponics or the nutrient film technique (NFT) may necessitate a deeper knowledge of hydroponic principles and procedures.

- *Maintenance Requirements:*

Hydroponic systems of different types exhibit differing levels of maintenance demands. Certain systems necessitate regular maintenance, including monitoring nutrient and pH levels, regulating water flow rates, and regularly cleansing the system. While some systems may demand less attention, they may nonetheless have distinct maintenance requirements. Take into account the level of commitment and dedication you are prepared to allocate towards the upkeep of your system.

- ***Environmental Considerations:***

Familiarize yourself with the specific environmental requirements of the plants you aim to cultivate. Optimal plant development in certain hydroponic systems may necessitate precise environmental controls, including regulation of temperature and humidity. Assess whether the solution you have selected can meet these criteria within the available area that is expanding.

- ***Expandability:***

Reflect on your future objectives pertaining to the magnitude and scope of your hydroponic enterprise. If you have intentions of increasing your production in the future, choose a system that can be readily expanded or altered. This will prevent the inconvenience and expense of replacing your entire system when your requirements change.To summarize, selecting the appropriate hydroponic system necessitates a meticulous evaluation of variables such as plant choice, spatial availability, expenses, proficiency level, upkeep demands, ecological issues, and expandability. By assessing these aspects in relation to your own requirements and objectives, you can make a well-informed decision and position yourself for prosperous hydroponic cultivation.

CHAPTER 5

Plant Selection in Hydroponics

The choice of plants is a critical factor in hydroponic farming, as different plant species possess diverse needs and capacities to thrive in hydroponic systems. Through careful selection of suitable plants for your hydroponic system, you may enhance productivity, improve resource utilization, and guarantee the success of your cultivation endeavors.

Factors to consider

➤ *Growth Habit and Size:*

Take into consideration the plant's growth habit and size while choosing crops for hydroponics. Certain plants, such as vine crops like tomatoes and cucumbers, necessitate vertical support systems and lots of space to facilitate their growth. In contrast, leafy greens and herbs have more condensed growing habits.

> *Nutrients requirement:*

Plant species exhibit diverse nutritional requirements, encompassing nitrogen, phosphorous, potassium, and micronutrients. Choose plants with comparable nutrient requirements for easy nutrition management and to prevent nutrient solution imbalances.

> *Temperature and Climate:*

Take into consideration the temperature and climate needs of the plant species you desire to cultivate. Certain plants flourish in hot, tropical environments, while others prefer colder conditions. Select plants that are very compatible with the specific environmental parameters of your hydroponic system.

> ### ➤ *Light Requirements:*

Evaluate the light needs of the plants you intend to grow and verify that your hydroponic system offers sufficient amounts of illumination. Leafy greens and herbs often necessitate lower levels of light intensity in comparison to fruiting crops such as tomatoes and peppers.

> ### ➤ *Water and Oxygen Requirement:*

Take into consideration the water and oxygen requirements of the specific plant species you plan to cultivate. Certain plants, such as lettuce and spinach, possess shallow root systems and want regular watering, but others, such as tomatoes and peppers, possess deeper root systems and may necessitate less frequent irrigation.

➢ *Growth Cycle and Harvest Time:*

Consider the growth cycle and harvest period of the plants you intend to cultivate. Certain crops have rapid growth cycles and can be harvested numerous times within a single growing season, while others necessitate longer durations to achieve maturity.

➢ *Market Demand and Profitability:*

Conduct research on the local market demand and consumer preferences to ascertain the crops that yield the highest profitability when grown hydroponically in your region. When choosing plant species, it is important to take into account considerations such as cost, shelf life, and market trends.

➢ *Disease Resistance:*

Opt for plant cultivars that exhibit resistance to prevalent hydroponic diseases and pests in order to decrease the likelihood of crop failure. Utilizing disease-resistant cultivars can diminish the necessity for chemical interventions and guarantee the well-being and productivity of your hydroponic crops.

Which plants do best in hydroponic systems?

- **Leafy greens:** Leafy greens, like lettuce, spinach, kale, arugula, and Swiss chard, are ideal for hydroponic farming because they have shallow root systems and develop quickly. Hydroponic farmers can cultivate these crops using nutrient film technique (NFT) systems, deep water culture (DWC) systems, or vertical towers.

- **Herbs:** Herbs, including basil, cilantro, parsley, mint, and oregano, are commonly grown hydroponically for culinary purposes. Cultivating herbs in hydroponic systems requires minimal space and enables year-round growth. This method ensures a steady and abundant supply of fresh and aromatic herbs for culinary purposes.

- **Tomatoes:** Farmers frequently cultivate tomatoes as fruiting crops in hydroponic systems due to their lucrative market value and their ability to thrive in regulated settings. Because of their ability to grow vertically and generate abundant harvests within confined areas, determinate tomato cultivars are highly suitable for hydroponic farming.

- **Peppers:** Hydroponics is widely used to grow peppers such as bell peppers, chili peppers, and sweet peppers. Peppers are favored because they can produce large quantities and maintain consistent quality when cultivated in controlled settings. Peppers thrive in hot, sunny environments and thrive in hydroponic solutions rich in nutrients.

- **Cucumbers:** Cucumbers are climbing plants that may be cultivated in hydroponic systems, allowing for vertical growth and optimizing space utilization, resulting in higher yields per square foot. In hydroponic conditions, cucumbers grow when provided with sufficient support structures and a steady supply of moisture.

- **Strawberries:** Strawberries can be cultivated hydroponically utilizing vertical towers, hanging baskets, or nutrient film technique (NFT) systems. Hydroponic strawberries generate abundant quantities of delectable and succulent berries and can be collected numerous times during the growing season.

- **Microgreens:** Microgreens are the juvenile, delicate seedlings of vegetables and herbs that are collected when they reach the cotyledon, or first true leaf stage. Microgreens can be cultivated hydroponically in shallow trays or containers filled with a growing medium, offering a highly nutritious supply of fresh greens suitable for salads and garnishes.

The success of hydroponic farming operations greatly depends on the careful selection of plants, as each plant species has distinct requirements and adaptations to hydroponic systems. Growers can enhance yields, mitigate risks, and optimize profitability in hydroponic agriculture by taking into account elements such as growth habits, nutrient requirements, environment, market demand, and disease resistance. By employing careful strategizing and meticulous attention to details, hydroponic farmers have the ability to raise a wide array of superior crops throughout the year, thereby supplying local markets and communities with fresh and nourishing goods.

CHAPTER 6

Types of Nutrients and Nutrient Solutions Used in Hydroponics

Effective nutrient control is crucial for achieving success in hydroponic gardening. Hydroponic systems enable plants to obtain their nutritional requirements by relying on a nutrient solution while they grow without the use of soil. This solution supplies vital minerals necessary for plant growth and maturation. Gaining a comprehensive understanding of the significance of nutrients and their efficient management is essential for maximizing plant vitality, development, and productivity in hydroponic systems.

Essential Nutrients for Hydroponics

Hydroponic plants need a variety of vital nutrients to ensure robust growth. There are two primary categories of nutrients for hydroponic plants: macronutrients and micronutrients.

- **Macro nutrients**

These are essential nutrients that plants need in significant amounts. The main macronutrients include:

- **Nitrogen (N):** is necessary for the synthesis of chlorophyll and the overall development of plants.

- **Phosphorus (P):** is required for the growth of roots and the production of flowers and fruits.

- **Potassium (K):** is crucial for the activation of enzymes, regulating water, and maintaining general plant health.

- **Calcium (Ca):** is essential for maintaining cell wall structure and promoting plant growth.

- **Magnesium (Mg):** is a vital constituent of chlorophyll and plays a crucial role in the process of photosynthesis.

- **Micro nutrients**

Micronutrients, also referred to as trace elements, are essential for plant health despite being needed in small quantities. Several essential micronutrients include:

- **Iron (Fe):** is necessary for the creation of chlorophyll and the functioning of enzymes.

- **Manganese (Mn)**: plays a crucial role in photosynthesis, the activation of enzymes, and the metabolism of nitrogen.

- **Zinc (Zn):** is essential for the activation of enzymes and the metabolism of carbohydrates.

- **Copper (Cu):** is required for the proper functioning of enzymes and the metabolic processes in plants.

- **Boron (B):** is essential for the formation of cell walls, the production of pollen, and the setting of fruit.

- **Molybdenum (Mo):** is essential for the processes of nitrogen fixation and enzyme activity.

Common hydroponics nutrient

solutions

Hydroponic nutrient solutions come in diverse formulas tailored to suit the specific requirements of various plant species and growth phases. Several prevalent varieties of hydroponic nutrient solutions include:

- **Pre-formulated solutions**: This refers to nutrient solutions that are readily available in the market. These solutions are designed to contain a well-balanced combination of vital elements in precise ratios. Pre-formulated solutions are convenient and user-friendly, making them suited for both novice and seasoned cultivators.

- **Two-part solutions:** This refers to nutrient solutions

that are composed of distinct bottles of liquid fertilizers containing the essential macronutrients and micronutrients. Growers have the ability to modify nutrient concentrations by blending the two components in varying proportions in order to satisfy the individual needs of their plants.

- **Custom Formulations:** Certain cultivators opt to produce personalized nutrient solutions by combining distinct nutrient salts based on the precise requirements of their crops. Custom formulations offer enhanced flexibility and control over nutrient amounts, albeit necessitating careful monitoring and adjustment.

Managing Nutrient Solutions in Hydroponics

Effective administration of the nutrient solution is crucial to guaranteeing that plants obtain the required nutrients in the appropriate proportions. Below are a few crucial factors to take into account when overseeing nutrient solutions in hydroponic systems:

- **pH Level:** The acidity or alkalinity of the nutrient solution impacts the accessibility of nutrients to plants. Plants generally thrive in a pH range of 5.5 to 6.5, which is somewhat acidic. Consistent monitoring and fine-tuning of pH levels are essential to avoid nutritional shortages or toxicities.

- **Nutrient Concentrations:** Monitoring nutrient concentrations is essential for preventing shortages or excessive levels. One can modify nutrient concentrations by incorporating suitable fertilizers or altering the dilution of the nutrient solution.

- **Electrical Conductivity (EC):** EC is a parameter used to quantify the level of dissolved salts in a nutritional solution. It quantifies the potency of the nutrient solution and aids in assessing the requirement for supplementary nutrients. Ensuring a consistent EC level is crucial for plants to efficiently absorb nutrients.

- **Water Quality:** The utilization of

uncontaminated, superior-quality water is vital for the formulation of nutrient solutions. Water supplies must be devoid of pollutants and pathogens that have the potential to damage plants or interfere with nutrient absorption.

- **Nutrient Solution Temperature:** The temperature of the fertilizer solution can influence the availability of nutrients and the health of the roots. For best nutrient uptake and root growth, it is ideal to maintain nutrient solutions within the temperature range of 18°C to 22°C.

CHAPTER 7

Troubleshooting Common Hydroponic Issues: Solutions for Optimal Growth

The popularity of hydroponic gardening has surged because of its resource efficiency, capacity to cultivate plants in small spaces, and superior yields in comparison to conventional soil-based techniques. Nevertheless, similar to any other form of cultivation, hydroponics is not devoid of its own set of obstacles. Cultivators frequently face diverse challenges that can hinder plant development and output. It is essential to comprehend and resolve these prevalent hydroponic problems in order to sustain vigorous plants and attain maximum yields.This guide will look at common problems that arise in hydroponic systems and provide practical strategies to efficiently resolve them.

> ## ➤ Nutrient Imbalance

An essential part of hydroponics is ensuring that plants receive proper nutrients in the appropriate ratios. Nutrient imbalances may occur as a result of factors such as inappropriate nutrient blending, variations in pH levels, or insufficient nutrient absorption by plants. Indications of nutritional imbalance encompass the occurrence of yellowing leaves, hindered growth, and leaf discoloration.

Solution:

- Consistently check nutrient levels by utilizing an electrical conductivity (EC) meter and a pH meter.

- Adhere to the manufacturer's instructions for the proportions of nutrients to be mixed, and make any required adjustments depending on the stage of plant growth.

- Regularly flush the system to avoid the accumulation of salt, which may interfere with the absorption of nutrients.

- Maintain the pH levels within the ideal range of 5.5 to 6.5 for individual plant species.

- If shortages are identified, it is advisable to utilize a

well-balanced nutrient solution or incorporate a micronutrient supplement.

➤ Root Rot

Pathogenic fungi and bacteria primarily cause root rot, a prevalent problem in hydroponic systems. It is particularly common in settings characterized by high humidity and inadequate oxygenation. Indications comprise mucilaginous, dark roots, unpleasant smells, and withering vegetation. If not handled, root rot can have a substantial influence on the absorption of nutrients and the overall health of plants.

Solution:

- Ensure enough oxygenation and aeration in the

hydroponic system by employing air stones, air pumps, or oxygen injectors.

- Ensure the water temperature remains within the ideal range of 65°F to 75°F or 18°C to 24°C to prevent the proliferation of pathogens.

- Employ sterilized growing media and utilize clean, sanitized equipment to minimize the entry of germs.

- Establish a proactive treatment protocol by employing advantageous bacteria and fungi to overcome detrimental diseases.

- It is advisable to include hydrogen peroxide or other natural antifungal medicines to manage the proliferation of microorganisms in the root area.

➢ Temperature Fluctuations

Temperature variations can exert a substantial influence on the development of plants, the absorption of nutrients, and the overall well-being of hydroponic systems. Severe temperatures can cause plants to experience stress, decrease the solubility of oxygen, and facilitate the growth of diseases and algae.

Solution:

- Ensure that the temperature in the growing environment is kept at an ideal level to promote the healthy growth and metabolic processes of plants.

- Employ temperature regulation devices such as heaters, chillers, or evaporative coolers to control water and air temperatures.

- Implement temperature sensors and monitoring systems to accurately measure temperature variations and promptly detect any problems.

- Ensure sufficient ventilation and air movement to disperse heat and maintain consistent temperatures across the entire cultivation area.

- It is advisable to utilize thermal insulation materials in order to maintain stable temperature levels and minimize energy usage.

➢ **pH Fluctuations**

Fluctuations in pH can have a substantial impact on the availability of nutrients and the ability of plants to absorb them. Various factors, including the concentration of nutrients, the quality of water, and the activity of microorganisms, can affect the pH levels in hydroponic systems. Deviation from the ideal range can cause nutrient lockout, a condition in which vital nutrients become inaccessible to plants, leading to nutrient shortages.

Solution:

- Consistently check the pH levels by utilizing a dependable pH meter, and make necessary adjustments by employing pH up or pH down solutions.

- Maintain a consistent nutrition solution and employ pH stabilizers, if needed, to regulate pH changes in the buffer.

- It is advisable to include pH buffering agents or additives in order to stabilize the pH values in the reservoir.

- Increase the frequency of monitoring and adjusting pH levels during periods of accelerated plant development or fluctuations in the environment.

➤ **Algae growth**

Hydroponic systems exposed to light, particularly in nutrient reservoirs and growing media, commonly face the issue of algae growth. Algae engage in competition with plants for nutrients and oxygen, resulting in diminished nutritional availability and heightened susceptibility to root rot. Excessive proliferation of algae can obstruct irrigation pipes and compromise the functioning of the system.Solution:

- Minimize the exposure of nutrient reservoirs and growing media to light by utilizing containers that are opaque or made of light-blocking materials.

- Ensure adequate water movement and aeration to reduce the presence of stagnant water and oxygenate the nutritional solution.

- To effectively manage the growth of algae without causing harm to plants, utilize algae inhibitors or algaecides that are formulated with natural components.

- Initiate frequent cleaning and maintenance procedures

to eliminate the accumulation of algae on system components.

- It is advisable to include UV sterilizers or ozone generators to prevent the formation of algae and ensure the maintenance of water quality.

Achieving success in hydroponic gardening necessitates extreme attention to detail and proactive control of any system-related concerns. To ensure ideal growing conditions and generate abundant harvests, gardeners can overcome frequent obstacles in hydroponic agriculture by comprehending them and employing suitable troubleshooting approaches. Hydroponic gardeners can optimize their indoor gardening efforts by swiftly and successfully managing nutrient imbalances, pH changes, temperature swings, root rot, and algae growth. By demonstrating dedication, patience, and steadfast adherence to optimal procedures, hydroponic enthusiasts can experience the gratification of flourishing plants and long-lasting sustainable production.

CHAPTER 8

Future of Hydroponics

The future of hydroponics is quite promising and holds significant potential as agriculture continues to advance. Below are a few projected trends and advancements:

Nutrient Innovation: Advancements in plant nutrition research will result in the creation of customized nutrient formulations designed to meet the specific requirements of certain crops. These compositions aim to enhance plant growth and production while avoiding waste and environmental harm.

Technological Integration: Hydroponic systems will progressively integrate automation, sensors, and data analytics to enhance plant growing conditions. This encompasses automated nutrient dosing, pH monitoring, and climate control systems to guarantee the most favorable development conditions.

LED Lighting: Ongoing advancements in LED technology will further enhance energy efficiency and the ability to precisely control the spectrum of light. This will allow gardeners to customize the light spectrum to meet the individual needs of different plants. LED illumination will become increasingly cost-effective and readily available for hydroponic cultivators.

Vertical farming: the practice of vertically stacking hydroponic systems, is expected to increase in popularity, particularly in metropolitan settings with limited space. Vertical farming optimizes land utilization and minimizes transportation expenses.

AI and Machine Learning Integration: The incorporation of artificial intelligence (AI) and machine learning algorithms will have a substantial impact on enhancing hydroponic operations. These technologies have the capability to evaluate extensive quantities of data in order to optimize growing conditions, forecast crop production, and identify and address potential problems such as pest outbreaks or nutrient deficits.

Consumer Demand for Fresh Produce: This is growing as customers place a higher priority on locally-grown fruits, vegetables, and herbs. This trend is expected to drive an increase in the market for hydroponically grown produce. Urban-proximate hydroponic farms will have the capacity to provide fresh produce throughout the year, hence diminishing the necessity for extensive transportation over great distances.

Sustainable techniques: With the rising awareness of environmental issues, hydroponic systems will progressively embrace sustainable techniques, including water recycling, the integration of renewable energy, and the utilization of organic inputs. There will be an increase in the prevalence of closed-loop systems that aim to limit resource usage and waste generation.

Expansion of Indoor Agriculture: The impact of climate change and unpredictable weather patterns on traditional farming is expected to drive the growth of indoor agriculture. Indoor hydroponic farming offers a dependable and climate-controlled alternative. This trend will result in the widespread establishment of expansive indoor hydroponic facilities.

In summary, the future projections for hydroponics are promising, as advancements in technology, sustainability, and efficiency propel the ongoing expansion and acceptance of soilless agricultural practices. Given the increasing urgency of concerns like population growth, food security, and environmental sustainability, hydroponics will have a progressively significant impact on the global food supply chain.

Acknowledgement

The Glory of this book's success goes to God Almighty and my ever-loving Family, Fans, Readers & well-wishers, Customers, and Friends for their endless support and encouragement.

ABOUT THE AUTHOR

Rita van Klaveren is an enthusiastic advocate in the field of agriculture and sustainable farming methods. Rita's extensive practical expertise in land management and crop cultivation has provided her with a profound comprehension of the intricate and demanding nature of contemporary agriculture. She embarked on her agricultural path at her family's farm, where she acquired an understanding of the importance of diligence, perseverance, and responsible management of the land. Rita's interest in hydroponics started during her undergraduate studies in environmental science, where she extensively explored the fundamentals of hydroponic systems and their capacity to transform conventional farming methods. Subsequently, she has committed her professional life to investigating the complex interplay of nutrient solutions, root-zone habitats, and regulated growing conditions. Rita's work, as a farmer and prolific writer, extends beyond the boundaries of academia, reaching farmers, entrepreneurs, and individuals around the globe. Through her perceptive publications and captivating lectures, she elucidates the intricacies of hydroponic farming, enabling people to utilize its revolutionary capacity. Rita's experience in hydroponics has enabled her to develop successful and sustainable systems for growing high-yield crops throughout the year, ranging from small urban gardens to huge commercial enterprises. Her methods prioritize environmental conservation. Her comprehensive strategy prioritizes sustainability, optimizing resource utilization, and promoting a wide range of crops, thus establishing a more resilient and food-secure future. Amidst a time characterized by increasing worries about the availability of food and the preservation of the environment, Rita emerges as a symbol of creativity and optimism, advocating for a concept of farming that goes beyond conventional limitations. With her steadfast commitment and limitless creativity, she persistently challenges the limits of hydroponic farming, leading the path towards a more environmentally friendly and bountiful future.